Britain's Geological H

by Robert Westwood

Inspiring Places Publishing
2 Down Lodge Close
Alderholt
Fordingbridge
SP6 3JA

ISBN 978-1-0682911-1-1

www.inspiringplaces.co.uk

The Green Bridge of Wales, Pembrokeshire

All photographs are by the author except page 83 by Ray Westwood, page 88 by Dr Alison Taylor, Wikimedia Commons and page 7 courtesy NASA.

Front cover: Norber Erratics, near Austwick, Yorkshire Dales.

Rear cover left to right: Gad Cliff, Dorset, Aysgarth Falls, Yorkshire Dales, Fingal's Cave, Isle of Staffa.

Robert Westwood is a writer/ photographer/publisher based in Dorset. He studied geology at the University of Exeter and has written several books about the Jurassic Coast, the geology of the Lake District and the Yorkshire Dales as well as walking guides for the Jurassic Coast Trust and the Cornish Mining World Heritage Site.

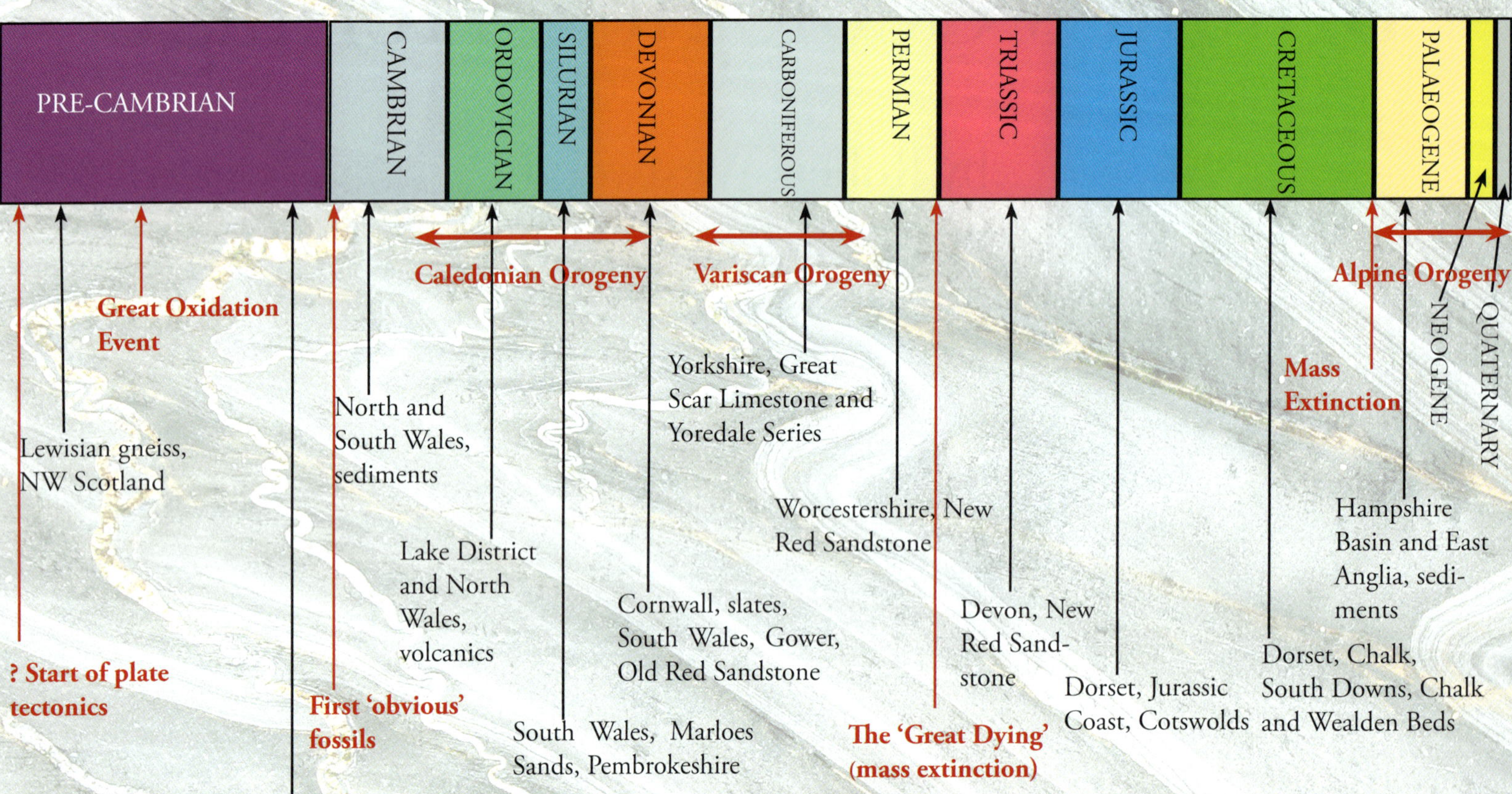

A timeline where, apart from the Pre-Cambrian, the width of the periods is proportional to their duration. It lists some of the formations that are mentioned in the book as well as some major geological events.

Contents

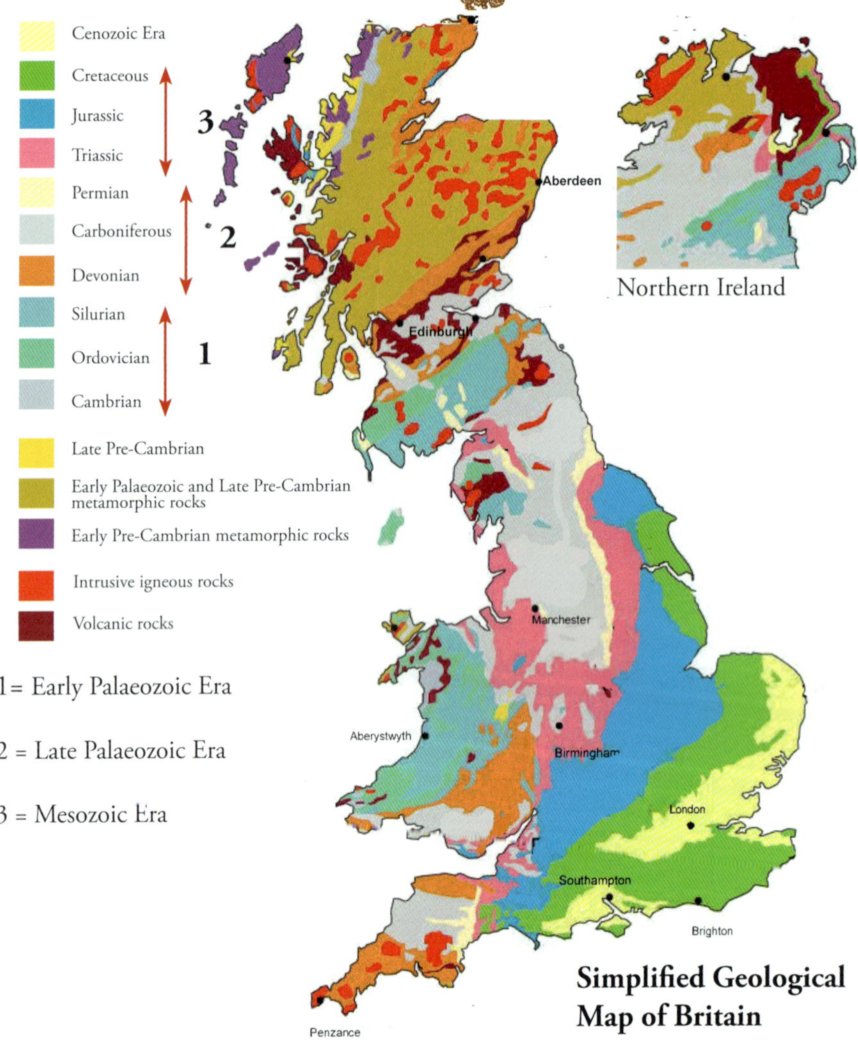

Simplified Geological Map of Britain

The Geological Timescale

Era	Period	Age my
Cenozoic	Quaternary	2.3
	Neogene	23
	Palaeogene	65
Mesozoic	Cretaceous	145
	Jurassic	199
	Triassic	251
Late Palaeozoic	Permian	299
	Carboniferous	359
	Devonian	416
Early Palaeozoic	Silurian	443
	Ordovician	488
	Cambrian	542

Opposite is a simplified geological map of Britain with a table showing the ages of the various geological periods. A quick look reveals a few facts about Britain's geology. Nearly all the rocks we see on the surface are younger than about 500 million years old; in very few places are there really ancient rocks. A few periods stand out as covering a lot of the country, specifically the Carboniferous, Triassic and Jurassic. It's also noticeable that the older rocks tend to be on the western side of Britain and the younger rocks towards the east.

Finally, it's worth noting what a variety of geology we have! This is only a small country yet rocks from all ages are represented here, rocks that have formed in many different environments. There's probably no better place to be a student of geology or simply an interested person who wants to understand a little about the multitude of processes that have shaped the Earth's outer layer.

Above: Britain's oldest rocks are in North West Scotland. This is the spectacular mountain Suilven. It's formed from Torridonian sandstone around a billion years old and rests on Lewisian Gneiss which is up to three billion years old.

Above: Now to some of Britain's youngest rocks. This is Christchurch Harbour, Dorset from Hengistbury Head which is composed of sands and clays from the Palaeogene Period less than 50 million years ago.

Introduction

Cherish – to love, protect and care for someone or something that is important to you (Cambridge English Dictionary). It should be obvious that the Earth is something very important to us and deserving of our care and protection, but it sometimes feels as if it's often taken for granted, that it will look after us rather than need looking after. Maybe it's because we don't appreciate how special it is, how many disparate processes have combined in its evolution and in the evolution of life. After all many, if not most, people would take care and appreciate something beautiful that has been exquisitely crafted. The aim of this book is to illustrate in words and pictures how some of these processes have worked to produce the marvellous, life-supporting outer layer of the planet we live on.

The atoms of tin in the pictured piece were forged in a previous generation of stars, perhaps in a low mass dying star (or stars) or, more exotically, in the merging of two neutron stars, themselves the collapsed cores of massive supergiant stars. This happened billions of years ago and perhaps, for millions of years, these atoms and those of other elements drifted in interstellar space before being drawn by gravity into a new star system. While a new star formed (our sun) other material collected into gaseous balls and fell into orbit around the star. Numerous collisions of matter generated heat, melting the rocky material that was accumulating. Gradually cooling and solidifying, planets formed, rocky planets nearest to the sun, giant gas planets further out. This was around 4.5 billion years ago; shortly after that it's thought that a massive impact with a small planet turned Earth into a 'magma ocean' and resulted in the formation of the moon. Heavier elements sank under gravity towards the centre, lighter ones concentrated in outer layers. Naturally, the outside of the planet cooled first, forming a crust around a still-molten middle. This is where something remarkable happened on our planet. No one is quite sure how, or exactly when, it started, but it seems that over three billion years ago the Earth's outer layer had broken into a number of 'plates' which were moving around relative to one another. There was no doubt about the source of energy to power this – the heat within the Earth provided ample – but the plates had to be strong enough to transmit force over thousands of kilometres, but not so strong that a solid, unbroken outer layer formed. We will talk more about 'plate tectonics' in a later chapter, but for now let's just marvel at the fact that this did start happening, for without it there would be no continents and we would almost certainly not be here.

Main picture: The Crab Nebula (courtesy NASA, ESA, CSA, STScI, Tea Temim (Princeton University)); this is a remnant of a supernova of a massive star. Heavier elements formed in such events; this was seen on Earth in 1054 CE.
Inset: A pendant from Blue Hills Tin, Cornwall, made from locally sourced alluvial tin (see next page).

Our tin atoms were now part of these plates (but only around 2 parts per million) that were slowly moving relative to one another. For several billion years they were probably 'recycled' many times, sometimes being part of a landmass, at other times perhaps in sediment at the bottom of the ocean. Around 300 million years ago they were caught up in a mountain building episode as two plates collided. As one (oceanic) plate sank underneath the other (continental) plate water released from minerals initiated melting of the overlying mantle and the sediments dragged down with it. The resulting liquid magma rose due to its buoyancy as far as it could and cooled slowly underground for thousands of years. Tin does not form minerals easily and so remained unattached as other elements joined together and did so. Eventually, when most of the magma had solidified (into granite) the temperature was low enough for the shy tin to combine with oxygen, of which there was still plenty. By this time groundwater was circulating around the still hot granite and some newly-formed minerals like the cassiterite our tin had formed were carried by these hot waters into cracks and fissures in the rock surrounding the granite intrusion. This is how many of Cornwall's mineral deposits were formed (copper is similarly unreactive).

Erosion gradually revealed isolated peaks of the huge granite 'batholith' that had formed. Rivers flowed from these granite uplands, cutting into the surrounding rocks on their way to the sea. The fall in sea level during the Ice Age resulted in an increase in the gradient of river valleys and hence faster, more powerful rivers, some of which eroded the mineral deposits. Cassiterite is quite a heavy mineral and was 'sorted' by the fast flowing rivers and concentrated at certain locations. These 'placer' or alluvial mineral deposits were the first to be worked in Cornwall, and were known and exploited in ancient times. The stream at Blue Hills was once a more powerful watercourse and has deposited eroded minerals along the valley. Cassiterite continues to be collected here and this is how our tin found itself in this charming piece.

Opposite page main picture: Tin mines at Botallack, Cornwall. These were once among the most profitable in Cornwall; tin production peaked in 1860. Steam technology allowed the mines to be extended around 400 metres out under the sea and about 500 metres deep.

Inset: The stream at Trevellas Coombe near St Agnes, Cornwall. This was a source of alluvial tin and is where the tin in the pictured piece came from.

What's been described is, of course, just a few of the multitude of processes that have shaped the continental crust we live on and provided raw materials which we have exploited. No human hand or mind has contributed to these processes but we are now lucky enough to be able to understand them to a certain degree. Perhaps if we marvel at and appreciate how physical laws and contingencies have led to our planet's outer layer teeming with life, we will be more disposed to treat it with respect and care. Looking at the story of tin in this light, how ridiculous it may seem that in the nineteenth century someone was said to own those tin deposits, enabling them to profit grotesquely from their exploitation. A greater understanding of the processes that have developed what we may call "natural resources" might perhaps foster a culture of more careful management of the Earth's crust and a more equitable distribution of its benefits: a distribution that includes all life, not just the human variety. One such natural resource is, of course, the land itself, the 'continental crust' that we all live on. This has not always been a feature of the Earth and its development is part of the story of plate tectonics and is the subject of chapter 4. First we'll have a look at what rocks are actually made of.

Opposite page: Kimmeridge Bay, Dorset.
This is the classic location for the Kimmeridge Clay, not that it is all clay but rather alternating layers of shale and limestones. These dark rocks were laid down in a warm, tropical Jurassic ocean where the bottom was anoxic or depleted in oxygen. This may have been because of the lack of circulation and it resulted in the accumulation of much organic matter (at Kimmeridge the rocks contain about 3.8% of organic matter). When deeply buried under the North Sea this organic matter has turned to oil and gas.

Over 2 billion years ago cyanobacteria began photosynthesising and producing oxygen. Once the iron in seawater had been oxidised and incorporated in sediment (the source of most of our iron and steel today) oxygen began to accumulate in the atmosphere. When organisms that photosynthesise die they decompose and use up oxygen; only rapid burial of organic matter can allow oxygen to accumulate and remain in the atmosphere. Without sediments like these at Kimmeridge, and ones that are currently forming in our oceans, life as we know it would not be possible.

Rocks and Minerals

The land we live on is part of what's known as the 'lithosphere', from the Greek word 'lithos' meaning stone. This is not just the Earth's crust but extends down a bit into the mantle, the layer below the crust. It ends in a zone known as the 'asthenosphere' between about 80 and 200 km below the surface. Due to the combination of temperature and pressure here the mantle behaves in a more ductile or fluid manner, so it's here that the 'rocky' zone ends.

Most people have probably learnt a bit about rocks at school, how there are three main types, igneous, sedimentary and metamorphic, and probably know a few examples like granite, limestone and slate. But what are rocks made of? That's an easy question to begin to answer; they're made of minerals, and a mineral is a naturally occurring chemical compound. Now we should go into a bit more detail. The two most abundant elements in the Earth's crust are oxygen and silicon, and most rock forming minerals are consequently silicates, that is, compounds of silicon and oxygen. The main exception is calcite (calcium carbonate) which is the main constituent of many limestones. This derives from organic activity, via organisms that make their hard parts like shells out of calcium carbonate.

Igneous and metamorphic rocks make up the vast bulk of the lithosphere and although here in Britain it's most likely that sedimentary rock is under our feet, remember these only form a relatively thin layer over parts of the surface, and basically apart from limestones, ultimately derive from weathered igneous and metamorphic rocks, most being deposited on the sea floor, particularly on the continental shelf adjoining the land. Metamorphic rocks are simply rocks that have been altered by heat and/or pressure. They may previously have been igneous or sedimentary. The main elements that join with silicon and oxygen are, naturally enough, ones that are relatively abundant. These are mainly aluminium, iron, calcium, sodium, magnesium and potassium. The most common group of minerals are the feldspars, making up over half of the Earth's crust. These are silicates of aluminium, calcium, sodium and potassium. Iron and magnesium silicates tend to be denser minerals and are concentrated in oceanic crust and the mantle.

Above right: A metamorphic rock known as schist on Ardnamurchan peninsular, Scotland. This was once fine grained sediment squeezed and heated during the collision of tectonic plates. The shiny nature is due to the mineral mica which formed from original constituent minerals as the temperature and pressure rose.

Above left: A close up of Cornish granite. The larger oblong crystals are feldspar; they crystallize first from the melt and so are able to fully form. The darker minerals are mainly biotite mica while quartz is the grey/colourless mineral that has crystallized last and so filled in the remaining space.

Bottom left: Typical Cornish slates at Constantine Bay. These were originally fine grained sediments which have been squeezed and relatively gently heated as they were caught up in a mountain building episode.

Although the chemistry of igneous and metamorphic rocks gets very complicated, one basic principle is helpful in explaining why we've got continental crust to live and walk on. When all the above elements (and more) are in a molten state, minerals rich in iron and magnesium tend to crystallize first. These are relatively dense minerals and sink to the bottom of the melt. Their removal changes the composition, the melt typically becoming more silica rich. It is also less dense and therefore more buoyant. This process is known as differentiation or fractional crystallization and can eventually lead to the formation of granite. This helps explain why continental crust, which is largely granitic in composition, can be thought to 'float' on the back of the tectonic plates and escapes being recycled at convergent plate margins (see page 18).

Left: Sedimentary strata at Sandymouth, Cornwall. These are sandstones and mudstones originally laid down in horizontal layers.

Opposite page: The valley of Watloes near Malham Cove in the Yorkshire Dales. The layers are mainly Carboniferous limestone laid down in a shallow, tropical sea over 320 million years ago. Limestone contains a high percentage of calcium carbonate, derived from the hard parts of living organisms such as those with shells. The chemistry is a little complicated but the precipitated calcium carbonate 'cements' the sediment often forming a hard rock that is useful as a building stone. Limestone is an indicator of a marine environment rich in life.

How do we know how old rocks are?

How do we know how old rocks are? Consider the work of historians and archaeologists; written records can provide accurate dates for events, but before that how do we know how old a Neolithic settlement is. Careful investigation can reveal the relative age of events and settlements, finding for example that an Iron Age hillfort was built on top of a Neolithic camp but it is techniques such as radiocarbon dating that give us absolute ages.

A similar process applies to geology and the age of rocks. The relative ages of strata can be worked out by careful mapping and with the use of 'zone fossils', fossils that only existed for a limited period of time and so mark certain horizons. This sort of work has been going on for many years and we now have a very accurate picture of the relative ages of the rock strata of Britain. It has only been possible to assign absolute ages to certain rock types since the advent of radio-isotope (or radiometric) dating, first invented by Ernest Rutherford in the early twentieth century. The radiocarbon dating mentioned above is similar in principle.

When igneous rocks cool and solidify they contain small traces of radioactive elements. Over time these decay by emitting particles at a definitive rate. This leads to the establishment of a 'half-life' for each element whereby it takes a certain time for half of the original amount of that element to decay. For example, half an amount of uranium-235 will decay to lead-207 in 700 million years. After 1400 million years, one quarter of the original amount will be left. By analysing the relative amounts of such radioactive elements an accurate estimation of the age of igneous rocks can be made.

Opposite page: The Ennerdale granophyre in the Lake District is an igneous intrusion, much the same as a granite but with smaller grain size. It has been dated using uranium - lead radiometric dating to 450 million years old, +/- 3/4 million years.
Inset: An ammonite on the shore at Kimmeridge, Dorset. Ammonites evolved quickly and make good zone fossils.

Moving Plates

The outer rocky layer of the Earth is divided into a number of 'plates' which move relative to one another.

It was in 1915 that Alfred Wegener coined a term that caused a great deal of controversy in the scientific world. That term was 'continental drift' and it was an apt title for his theory that the Earth's continental landmasses move slowly relative to one another, driven by forces that, over geological time, create new continents and break existing ones apart. It was pointed out that it was no coincidence that the shapes of Africa and South America seem to fit together like a jigsaw puzzle, they once had done.

There was plenty of opposition to this seemingly outrageous suggestion, largely based on the assumption that if geologists and other scientists had no idea how continents could move around the surface of the Earth, then they could not do so. Soon, however, the evidence accumulated that this had happened and still does. There is not space to go into that evidence here but suffice to say that it is pretty unequivocal, and that was before modern instruments that can actually track and measure the movement of continents. It turns out that it is in the order of a few centimetres per year, commonly approximated to about the rate your fingernails grow.

Today a lot of research focusses on how and why the phenomenon of 'plate tectonics' shapes the outer layer of our planet. After a somewhat brief discussion about the mechanism we will see how it explains many aspects of Britain's geology and how we can see direct evidence of it all over our landscape.

Most oceanic crust is less than 200 million years old; this is due to the fact that it is being continually recycled. New oceanic crust is produced at mid-oceanic ridges, fissures down the centre of the oceans where new material in the form of basalt lava is constantly rising. At the other side of many plates the oceanic crust is being subducted, or drawn down into the mantle (see diagram page 22). When the magma first erupted it was hot and relatively buoyant, as it moves further away from the mid-ocean ridge it cools and becomes more dense. This causes it to sink back into the mantle

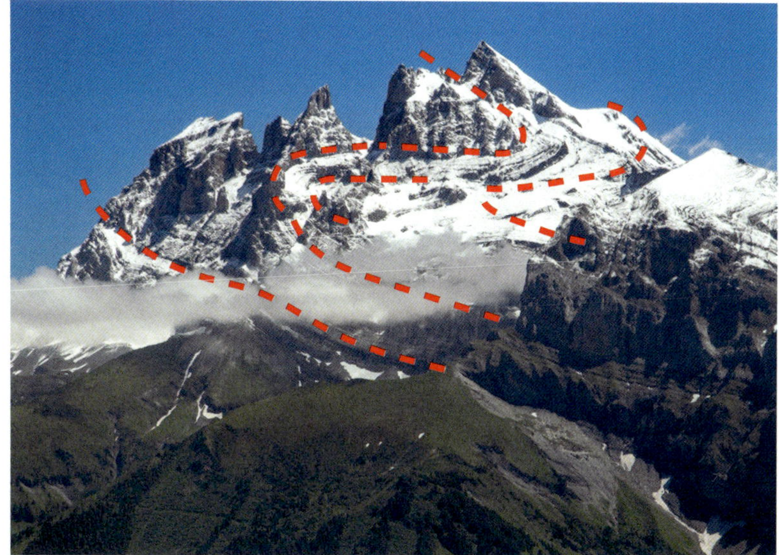

Top left: The Lake District - the core of this beautiful region is composed of volcanic lavas, tuff and ash from the volcanoes of an 'island arc' system, a chain of volcanic islands where one oceanic plate is subducting under another.

Top right: Fold mountain chains are formed when two tectonic plates collide, either two continental plates or a continental and oceanic plate. This is the Dents du Midi, France, the red lines show that intense folding has led to 'recumbent' folds called 'nappes', the main pressure has been from right to left (south to north as the African plate collided with the European).

Recumbent folds in the cliffs of Crackington Haven, Cornwall; the core of a once great mountain chain.

when it meets another plate (a destructive plate margin). It is the drag from this subduction that largely drives plate movement, causing the cracks or fissures that are the mid-ocean ridges. Here the release of pressure causes the mantle beneath to partially melt, giving rise to the volcanic activity all along the ridges (Iceland sits on top of the Mid-Atlantic Ridge). You might immediately ask yourself how, if it is the drag from sinking plates that is the main cause of plate movement, did this start in the first place – it seems a chicken and egg dilemma. There is still no definitive answer to this question; there is very little original continental crust left and no original oceanic crust. Also conditions 3 billion years ago were very different from the present; the mantle would have been hotter and perhaps thinner, weaker crust was more easily broken up by rising hot plumes. However, for plate tectonics to happen, the plates still had to be strong enough to transmit pressure over long distances. The only consensus among scientists is that plate tectonics has evolved to its present format, it may have been quite different when it was initiated.

Perhaps another obvious question is what do the plates move on? Are they disconnected from the mantle beneath? This is easier to answer – a little way down in the mantle, at the bottom of what we call the lithosphere, is a region known as the asthenosphere. This layer is found between 80 and 200 kilometres below the surface and is around 500 kilometres thick. It is a region where the temperature and pressure combine to make the mantle rocks more ductile or mechanically weak. Below this layer, although the temperature increases, the pressure is too great to allow such softening of the rocks. The ability of the asthenosphere to flow allows the movement of the rigid plates which 'float' above it, a movement as we have seen that is measured in centimetres per year. The rock in the asthenosphere is near melting point and this helps explain why, at mid-oceanic ridges where the plates are being pulled apart, the release of pressure allows the rock to fully melt and erupt on the ocean floor.

We can now look at the landscape with the knowledge that it is not static, that continents have moved around for millions of years. As Alfred Lord Tennyson observed:

> "They melt like mist, the solid lands,
> Like clouds they shape themselves and go." (In Memoriam).

Highly contorted slates at Boscastle, Cornwall. These were once fine grained sediments; heat and pressure turned them to slate and further deformations produced these wonderful fold patterns.

Those areas that are on the boundary of plate collisions are unsurprisingly areas of often violent tectonic activity – earthquakes and volcanic eruptions. Other areas in the middle of plates are relatively safe, quiet and stable. This is the situation we in Britain find ourselves in, no active plate boundaries are anywhere near us. However, a number of times in the past this land has been in such unstable areas and although millions of years of erosion has worn away and rounded once mighty mountains there are many features of our landscape that make sense once we understand the role of plate tectonics. We will see many such features as we look at the various ages of Britain's geological heritage and later pages will look at a few relevant examples.

It's probably obvious that the geological features we will be looking at are part of the continental crust, and it should be pointed out that the existence of continental crust, the land we live on, is itself a product of the movement of the plates and is intimately associated with the creation of mountain ranges. The next chapter looks at how continental crust is formed and features Britain's oldest rocks, formed at a time when there was far less continental crust on the Earth. Chapter 5 shows how various geological features are explained by plate tectonics.

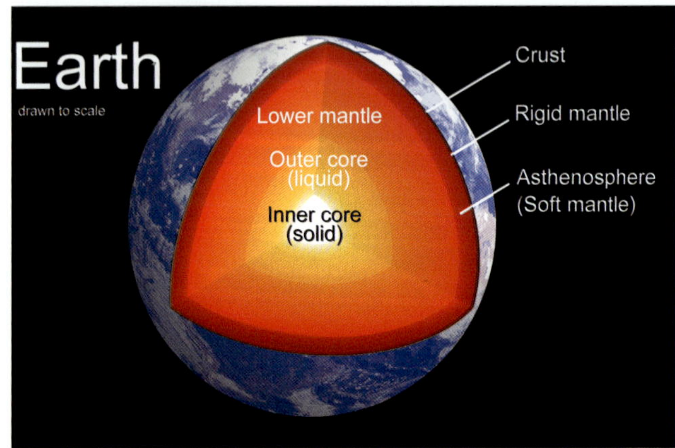

A schematic diagram through the Earth. Courtesy IsadoraofIbiza.

Opposite page: Lake District Volcanics - The core of the Lake District is largely composed of Ordovician volcanic rocks, lavas, ashes and tuffs. These rocks were formed in a relatively short time, about 5 million years. When an oceanic plate collides with a continental plate or when two oceanic plates collide, the denser oceanic plate will sink back into the mantle. Seawater locked in the plate's minerals will be released and will lower the melting point of mantle material. As molten magma rises this will give rise to volcanic activity above the sinking plate, often forming a volcanic island arc, a series of volcanic islands along the line of the subduction zone. The picture shows the fells above St John's in the Vale; the harder layers of lava give the stepped appearance of the hillside.

The Formation of Continental Crust

Some of the very first continental crust is exposed in North West Scotland.

The outer layer of the Earth is called the crust; its relative thickness is often compared to the skin of an apple and occupies about 1% of the Earth's volume. There are two types of crust, continental and oceanic, their names accurately indicating their distribution. It is worth just listing some facts about the crust, for the explanations of these facts help us understand many of the geological processes that have shaped this surface layer of the planet.

Continental crust is between 30-50 kilometres thick.
Oceanic crust is between 5-10 kilometres thick.
The oldest continental crust is over 4 billion years old.
The oldest oceanic crust is about 250 million years old.

Opposite page: Achmelvich
The tiny coastal village of Achmelvich sits on the three billion year old rocks known collectively as the Lewisian Gneiss. Most of these rocks were originally igneous but they have been repeatedly altered deep within the crust. Two main episodes of intense deformation affecting these rocks have been identified, one around 2.5 billion years ago and a later one, a prolonged mountain building event, around 1.9 to 1 billion years ago. These mountains were among the first to appear on our planet, although plate tectonics had been happening for some time. It's difficult to imagine what they would have looked like with no vegetation covering but safe to assume that they would have been eroded much more quickly than the mountains we see today. There has been recent research that suggests the formation of these early mountains depended on the abundance of life in the oceans. The proliferation of cyanobacteria and their subsequent burial led to ocean sediments being rich in carbon. This is thought to have helped 'lubricate' the layers of sediment as they were deformed and crushed by movement of the plates, allowing the crust to thicken and form mountain chains. Here at Achmelvich we can walk on the eroded core of these ancient mountains.

Sutherland, NW Scotland. The area known as Assynt has a spectacular landscape of isolated mountains sitting on some of the oldest rocks on the planet.

First the thickness disparity. Oceanic crust is denser than continental and if we think of the crust as 'floating' on the layer beneath (the mantle) then the thickness and density differences balance out – i.e. two columns of rock from a continent and the ocean, from the surface to the centre, would have roughly the same mass.

Now the age; we saw in the chapter on plate tectonics that mid-ocean ridges are where new oceanic crust is being formed and that it is at the destructive margins of the plates – subduction zones – where the oceanic crust returns to the mantle. Thus, as the plates are continually moving, the oceanic crust can only exist in the time it takes for it to be recycled. Continental crust, on the other hand, is formed at the collision zone of plates. Partial melting of oceanic plates and subsequent 'differentiation' of the melt (see also page 14) leads to the formation of lighter rock. This tends to ride on the back of plates and is not usually subducted, although of course the land is gradually eroded, detritus returned to the sea and may be subducted in due course or squeezed up into a mountain chain at a destructive plate boundary. Thus the land we live on is the product of the movement of the plates over time. The rocks we see on the surface will usually be relatively young, often being the product of oceans and seas that have been crushed out of existence or from times when the sea level was higher (or the land lower!). They may have been formed in rivers, deltas or estuaries or in windswept deserts. There may be volcanic rocks that have been extruded or emplaced in association with plate movement. Underneath this 'surface cover' of relatively young rocks is what is known as the 'basement', older rocks that are metamorphic or igneous in nature, often very complex and highly metamorphosed.

Opposite page: Achnahaird Bay, NW Scotland.
Achnahaird Bay is also cut out of the Lewisian Gneiss, but here, in the background, we see the mighty 'inselberg' mountains formed from the Torridonian sandstone which is roughly 1 billion years old. The term 'inselberg' simply means an isolated mountain, they are the remnants left after the erosion of many thousands of feet of the sandstone layers. Intriguingly it is thought the sea level at the time these sandstones were first being deposited was roughly where it is today. They were laid down, on the already eroded core of the 'Lewisian mountains' mostly by rivers and perhaps shallow lakes in a basin which slowly subsided as the layers formed.

This leads to more questions though; is the amount of continental crust increasing or is it gradually being depleted by erosion and subsequent subduction? In fact it appears that the amount of continental crust is quite stable, which leads us to assume that in the past its formation was much quicker, leading to the building of the continental masses we see today. There is evidence that some sort of plate tectonics had begun around 3.6 billion years ago, and that quite possibly the Earth was a 'water world' around 3 billion years ago. This may have been due to there being little continental crust at that time and/or there being more water.

As you might expect, the ancient basement rocks appear only rarely at the surface except in areas known as 'cratons' which sit at the centre of stable continental land masses (e.g. parts of Africa and Canada). However, here in Britain we have our own outcrop of very old basement rocks in the north west of Scotland and the Outer Hebrides, in fact, some of the Earth's earliest continental crust. Known as the Lewisian Gneiss (after the Isle of Lewis) it is a metamorphic rock, a rock that has been altered from its original state to something different, usually by heat and/or pressure. It's not usually that the overall chemical composition changes but rather that new minerals form from the elements present and that the structure of the rock changes. If the temperature rises so that the original rock melts then we can say a new igneous or volcanic rock has formed when it cools and solidifies. Much metamorphism happens deep in the crust as the collision of plates gradually squeezes large volumes of rock caught between.

Gneiss is known as a 'high grade' metamorphic rock, formed at high temperature (greater than 300 degrees Celsius) and pressure, typically at depths of 10-30 kilometres. This is enough to make the rock 'plastic', to enable it to flow and deform and for new crystals to form. It is usually associated with a mountain building event. A frequent characteristic

Opposite page: Quinag, one of the inselberg or isolated mountains of Assynt. Like the other peaks in the area it is composed of ancient Torridonian sandstone and sits unconformably on top of the Lewisian Gneiss. Incredibly, the gneiss is up to 3 billion years old and the mountain chain it was part of had been worn almost flat by the time the rivers and lakes deposited the sandstone around 1 billion years ago! Ice sheets scoured this landscape just 10-15 thousand years ago.

of gneiss is the banding of light and dark minerals. It can form from different original rock types; the Lewisian Gneiss derives mainly from igneous rocks intruded deep within the Earth's crust. It did not form in a single event but represents a complex history of igneous intrusion and intense metamorphism and deformation. Much of the deformation and banding was formed by metamorphism around 1.7 billion years ago.

To the left and above are pictures of the Lewisian Gneiss from the coastline of Assynt. All of them show the characteristic banding of light and dark minerals. The picture above left also shows some of the 'flow' structures, indicating the rock behaved in a semi-fluid or plastic manner as it was deformed by the intense heat and pressure.

Opposite page: Bel Tor, Dartmoor. The upper continental crust is basically granitic in composition. The Dartmoor Granite is part of a huge body of granite called a 'batholith' that underlies much of South West England. It was formed towards the end of the Variscan Orogeny (see page 74) as two continental plates collided. It derives from the melting of earlier sediments that had collected in ocean basins between the converging plates.

Some Features Explained by Plate Tectonics

Folded rock strata - *how is it that layers of solid rock can be folded and contorted as if they were soft putty?*

I've long been fascinated by folded layers of rock strata. They present in such a wide variety of forms: smooth, elegant and symmetric, angular and asymmetric, tightly crammed together like a concertina and sometimes so convoluted they defy description. They come in all sorts of sizes – some are easily visible across a whole mountainside while others span just a few centimetres.

It might be an obvious thing to say but for folded layers or strata to be visible there must be different layers of rock to begin with. When we see folded rock strata they will almost always be folded layers of sedimentary rock; the layers represent different types of sediment and/or different environments where the sediments were deposited. For example, a rise in sea level might result in sands of an estuary being covered in finer mud or clays. Many of the sedimentary rocks we see will have been deposited in a marine environment and will usually have originally been laid down in horizontal layers.

As sediment builds up increased pressure and chemical changes turn the soft sediment to solid rock. What then can bend and fold these layers into the contorted shapes we see in cliffs and hillsides? The answer, of course, is the collision of tectonic plates. But this is only part of the explanation; rock like limestone or sandstone is brittle, put between a powerful vice such rock would fracture and shatter. The folding of strata is often likened to strips of play-dough being squeezed between blocks of wood, but how can rocks behave like this?

The answer lies in the increased temperature and pressure within the continental crust. It's thought that rocks begin to behave in a ductile manner, that is they begin to deform without presenting macroscopic fractures, at depths of between 10-15 kilometres. Any deformation takes place very slowly, remember, the tectonic plates move at speeds

This spectacular fold is in Upper Carboniferous strata at Saundersfoot, Pembrokeshire, South Wales.

measured in a few centimetres per year. As you might expect, the nature of the rocks partly governs the types of folding produced, different strata have different properties. When we see contorted and folded strata the rocks have undergone viscous deformation, that is they have behaved more like a fluid. This happens at very high pressures and temperatures.

The more intensely folded strata are to be found largely in the western parts of Britain, in areas that were once associated with mountain building and the collision of the tectonic plates. But compressional forces were felt widely in areas more distant from the intense deformation. The trend of folds tells us the direction of the compression; push a sheet of paper from both ends, say from the north and south, and the trend or 'axis' of the resulting fold will be east to west. Thus the trend of the majority of major folds in South West England resulting from the Armorican or Variscan mountain building episode is east to west. This tells us that the compression was in a north-south direction. Further away from the centres of intense deformation stresses may be transmitted that cause more gentle folds. Consider the simple geological map of South East England; here the layers of Cretaceous sediments, largely Chalk and sandstones have been folded into a very broad anticline, with the older sediments in the middle since the overlying layers have been eroded away. The stresses that caused this folding came from the collision of the African and European plates. This led to the formation of the Alpine mountain range over millions of years, beginning around 55 million years ago. The force to buckle the rocks of the Weald may have been transmitted through older, hard, 'basement' rocks below more recent sediments.

Some rocks may go through a number of episodes of deformation. The movement of the plates of the Earth's lithosphere varies over time; a particular region may be subject to compression at one time which over thousands or even millions of years may change to a shearing or extensional movement. This can produce complicated fold patterns.

The evidence that rocks have been folded does not always present in wonderful contorted shapes. Sedimentary layers laid down on the sea floor are typically horizontally aligned, when we see such layers lying vertically or steeply inclined

This is the South Downs escarpment near Washington, Sussex. The hard Chalk and Upper Greensand below dip gently southwards (to the left on picture), part of the southern limb of the anticlinal fold. On the North Downs these beds dip gently northwards. Between them lie the soft sands and clays of the Wealden Beds (see page 88).

we know they have been subjected to powerful earth movements. They may simply be the limbs of large scale folds where the curved parts or hinges have been eroded away. The escarpments of the North and South Downs are the eroded tops of the limbs of the broad anticlinal fold of the Weald formed from the more resistant Chalk and Greensand.

Crustal shortening - When tectonic plates collide any compressional forces will typically act in a horizontal plane (ignoring the curvature of the Earth) and the crust caught between the colliding plates will be squeezed and shortened. Sometimes the folded layers of sedimentary strata can be used to estimate the amount of crustal shortening. Often, of course, the folding is rather complicated but sometimes the folds are simple enough and distinct enough for anyone to gain a rough idea of the amount of shortening. Consider the two examples in the pictures, one of chevron folding in the cliffs at Millook Haven, Cornwall, the other, as we have seen, at Saundersfoot in Pembrokeshire. By doing a very rough and ready calculation (c/(a+b)) we can work out an approximation for the crustal shortening that these folds represent. It turns out to be in the region of 60% shortening for Millook Haven and around 40-50% for Saundersfoot. More accurate measurements have been done and these figures are reasonably accurate. These folds date from the same time, the Variscan mountain building episode (see page 74), and indicate that Cornwall was nearer the centre of deformation than South Wales.

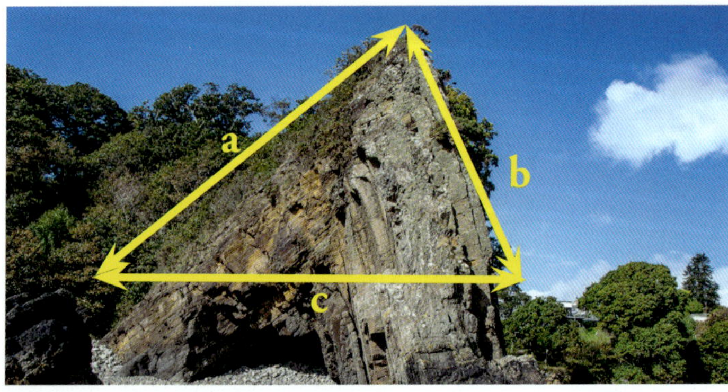

Millook Haven

Porth Clais on the Pembrokeshire coast, South Wales. These Cambrian sandstones have been folded during the Caledonian Orogeny. We now see the layers lying almost vertically.

Faults and thrusts - *fractures in rocks are the result of brittle deformation.*

Folded and deformed rocks are typically the result of compression caused by the movement and collision of the plates of the Earth's crust and upper mantle. We have seen how, under pressure and relatively high temperatures, rock strata behave in a fluid, ductile manner. When strata are not under such pressure, that is when they are nearer the surface, they will tend to fracture rather than flow. The movement of the plates on a spherical surface is complicated and their relative movements can result in tensional forces as well as compressional. It is tensional forces that result in 'normal' faults where one crustal block slips down relative to another. Compression can result in thrusts where a section of crust is thrust over an adjacent section.

Some faults and thrusts are major, large scale features, others are very much smaller. Some tensional forces are caused by strata being uplifted, perhaps those overlying an igneous intrusion which rises due to its natural buoyancy. On a very small scale this may result in joints in the overlying rocks, small fractures where there is no obvious vertical or horizontal displacement.

The evidence for these large and small scale structures is all around and we will look at some examples which illustrate the forces involved and give further insight into the dynamic nature of our planet's outer layer.

Opposite page: Loch Linnhe in the Great Glen. This famous glen which includes a number of lochs including Loch Ness, follows the line of the Great Glen Fault, a 'strike-slip' fault or one where the relative movement was horizontal, the north side moving north-east and the south moving South West. It's thought the displacement was over 60 kilometres and formed towards the end of the Caledonian Orogeny (see page 62) towards the end of the Silurian Period and the beginning of the Devonian.

This picture was taken on Beer Head in East Devon, looking eastwards towards Seaton. We can see the cliffs of the familiar Cretaceous Chalk nearest while in the distance are cliffs of much older Triassic sandstone. Their juxtaposition is explained by a fault in between the two which has thrown the younger Chalk downwards so it is level with the Triassic sandstones. The fault is obviously younger than the Chalk and was probably caused by earth movements which formed the Alpine mountain chain.

This is Hope's Nose on the South Devon coast. The rock is the charmingly named Daddyhole Limestone of Devonian age. You can see the layers have been folded into a recumbent fold, formed by compression during a collision of tectonic plates during the Variscan Orogeny. The compression was in a north-south orientation with the 'push' from south to north (right to left of the photograph). Later on this also caused here a low angle thrust which can be seen running roughly horizontally in the middle of the photograph.

Unconformities

The movement of plates pushes up mountain chains, closes oceans and, more prosaically, uplifts and depresses areas of the Earth's surface. Regions that were land have been thrown below sea level while ocean floors have been elevated above. Climate changes have also been a major factor influencing sea level, particularly glacial episodes. Once above sea level the surface is subject to erosion; if the sea subsequently returns to this area, newer sedimentary layers may form on top of the old land surface. The result is an unconformity, sometimes the rock layers of different ages will be at an angle to each other, making the unconformity visibly obvious.

Below: The "De la Beche Unconformity" at Vallis Vale, Somerset, named after the famous geologist Henry De la Beche, the first director of the Geological Survey of Great Britain. Here flat lying Jurassic limestone sits on top of tilted Carboniferous limestone deposited around 340 million years ago. There is a time difference of about 170 million years between the two limestone deposits.

Below: Thornton Force in the Yorkshire Dales. This time it's horizontal layers of Carboniferous limestone lying unconformably on contorted Ordovician slates deposited around 450-480 million years ago, a time difference of over 100 million years. These slates were part of a great mountain range, eroded flat before the Carboniferous seas deposited the limestone.

Slate - *fine grained sediments crushed between plates.*

We're all familiar with slate; for centuries it has been used for roof tiles, and more recently for snooker tables and kitchen worktops. The properties that make it so useful are fairly obvious, it's relatively hard and splits easily into flat layers – and it is these properties that give a clue to its origin.

The main constituents of slate are clay minerals. Clays are the product of weathering of minerals in igneous or volcanic rocks, chiefly feldspars, and are hydrous aluminium silicates, often with other metals such as iron, magnesium, sodium and potassium. They are known as sheet silicates because they form flat, two dimensional crystals.

When two continental plates collide, or an oceanic plate subducts under a continental plate, huge volumes of sediments caught in between are subjected to increased temperature and pressure. This is known as regional metamorphism and the temperature and pressure rise as the depth of burial increases. What's known as 'low grade' regional metamorphism occurs at depths of a few kilometres and about 200 degrees Celsius. This increased temperature and pressure forces the clay minerals to slowly recrystallize (or regrow) and they do so at right angles to the direction of pressure, resulting in the flat crystals being aligned in one direction. It is this that gives slate its 'slaty cleavage', a propensity to split in a preferred direction. The pressure also compacts the rock even further, making it relatively hard.

So, this common, ordinary rock is evidence of the collision of tectonic plates and of collisions that formed great mountain ranges. We see slates on the beaches of Cornwall, in North Wales and the Lake District, often with the cleavages (preferred direction of splitting) aligned vertically. This is not surprising since it has been formed by pressure in the horizontal plane. Sometimes we can see that the cleavage has itself been folded and contorted by further episodes of deformation.

Opposite page: Devonian slates at Treyarnon, Cornwall. You can see the cleavage planes lie almost vertical.

Above: These small scale folds and shears are in Devonian slate at Boscastle Harbour, Cornwall. These were originally fine grained mudstones that underwent metamorphism and then subsequent deformation in the Variscan Orogeny (see page 74).

Right: Honister Pass in the Lake District. The high quality slate here has been mined for centuries. Likewise the slate from Tilberthwaite Quarry also in the Lake District (picture opposite page). Both slates have a fairly unusual origin - the fine grained sediment that was crushed by the collision of plates came from a volcano that was part of an island arc system. Huge quantities of ash from the eruptions fell in the shallow waters on the edge of the ocean that was being squeezed out of existence. It was this that was turned to slate by the collision of a continental plate with an oceanic plate.

Snapshots of Britain's Geological History

A look at various locations around Britain where rocks of different ages outcrop, beginning with the oldest.

The Pre-Cambrian > 542 million years BP

Pre-Cambrian is a label given to all rocks older than those of Cambrian age which began around 540 million years ago. Since the oldest known rocks are about 3.8 billion years old, this term covers most of the Earth's history. The Cambrian is significant because it is the earliest period from which easily identifiable fossils are found. Subsequent periods are defined by the fossil content found in their rocks.

By now you will be familiar with the fact that plate tectonics has shaped and re-shaped the Earth's crust over several billion years so it should perhaps be no surprise that Pre-Cambrian rocks are relatively rare, in Britain they occur predominantly in North West Scotland and in a small number of isolated locations elsewhere. Again, perhaps unsurprisingly, they tend to be rather hard rocks, igneous or metamorphic rocks associated with plate collisions and mountain building episodes or sedimentary strata that have been crushed by such episodes, an exception being the Torridonian sandstones of North West Scotland.

We have already met some of the Pre-Cambrian rocks of North West Scotland, the Lewisian Gneiss and the Torridonian strata. In England and Wales there are much smaller outcrops in Charnwood Forest, near Nuneaton, Shropshire, The Malvern Hills, South Wales and Anglesey and the Lleyn Peninsula.

North West Scotland

Just to recap – the Lewisian Gneiss which forms the base of much of the Assynt region, and, of course, the Isle of Lewis, is a metamorphic rock that was mostly originally igneous. The original rocks are up to 3 billion years old and

Couldn't resist another photo from Assynt, North West Scotland. This is the iconic mountain Suilven (and inset from summit), formed of Pre-Cambrian Torridonian sandstone.

have been crushed, deformed and baked several times to leave the gneiss we see today (see page 30). They were once part of a great mountain range, eroded almost flat before around one billion years ago rivers and lakes in an arid environment began to deposit the Torridonian sandstones that today form the iconic mountains of Assynt. Thousands of feet of these sandstones have been worn away to leave the dramatic isolated peaks and reveal the gneiss underneath.

Sometime during this period, now thought to be about 990 million years ago, a calamitous event occurred which has left tantalising traces in the rocks. A particular stratum known as the Stac Fada Member outcrops on the coast near the village of Stoer on the north west coast. Originally thought to be some sort of volcanic deposit, recent research has uncovered evidence that suggests it records the impact of a giant meteorite. In particular it is the presence of grains of quartz that show signs of a great shock that convinced scientists it was part of the outward flow of material from the impact. This has now been refined to suggest it was formed by sediment being flung outwards by the force of the impact and not directly material displaced. The location of the site of the impact is still unclear.

Between the Great Glen and the Lewisian rocks of the far north west lie rocks of what is known as the 'Moine Supergroup'. These were originally sands and muds deposited in a fairly shallow sea on the eroded surface of the much older Lewisian rocks between about 1000 and 800 million years ago. They were subsequently deformed and metamorphosed in several major tectonic events, culminating in the Caledonian Orogeny in the Palaeozoic Era. As well as the famous Moine Thrust there are a number of other thrust faults, all trending north-west to south-east and indicating a 'push' from south-east to north-west.

Right: The Moine Thrust is clearly visible at Knockan Crag. The Pre-Cambrian Moine rocks have been thrust over the older Lewisian rocks, from right to left in the picture.

The Stac Fada Member near Stoer, North West Scotland. This is thought to be part of the 'ejecta blanket' thrown out after a meteor impact around 990 million years ago. The large, light coloured roundish block in the centre may be Lewisian Gneiss torn up and carried by the flow similar to small chunks in the inset picture.

Shropshire

In Shropshire lie rocks of so called 'Uriconian' and 'Longmyndian' age, what we might call young Pre-Cambrian just before the onset of the Cambrian Period. There are deformed and metamorphosed sediments and volcanics thought to have been formed in an island arc situation where two oceanic plates collided, eventually closing the ocean and forming a mountain chain. The subduction of one plate initiated melting in the crust leading to the volcanic activity. On the east side of the Church Stretton Valley lie the volcanic hills of Caer Caradoc, The Lawley and The Wrekin while to the west lie the sediments which form the Long Mynd. The valley follows the line of the Church Stretton Fault which became active during the tectonic activity and remained active for many millions of years. The sediments of the Long Mynd are thought to have derived from the mountains formed by the plate collision after a relatively short time geologically. They are highly deformed, caught in a tectonic vice as the ocean continued to close.

Far left: Longmyndian strata in Cardingmill Valley near Church Stretton. These are sediments from the late Pre-Cambrian that have been crushed and deformed as two plates collided.
Left: Hard, volcanic lavas on the summit of Caer Caradoc.

Opposite page: The picture shows the valley which marks the line of the Church Stretton Fault. This fault runs north-east to south-west through Shropshire and into South Wales and is a strike-slip or tear fault similar to the Great Glen Fault. On the far side are the hills of Caer Caradoc and The Lawley, formed of Pre-Cambrian volcanic rocks that are thought to have been formed when oceanic and continental plates collided about 570-550 million years ago. It is this collision which gave rise to the fault which was active over many millions of years. The bottom of the photo shows the high ground on the other side of the valley composed of folded sedimentary rocks, again Pre-Cambrian but younger than the volcanics.

The Malvern Hills

The Malvern Hills form a narrow ridge about twelve kilometres long, running north to south. Their relief is due to the hard, resistant rock of which they are formed, mainly plutonic igneous rocks that have been substantially metamorphosed. These rocks were formed nearly 700 million years ago above a subduction zone where one tectonic plate was sinking under another. Water carried down in the minerals of the sinking oceanic plate reduces the melting point of the mantle resulting in the formation of large bodies of magma which, being relatively buoyant, rise through faults and fissures creating a chain of volcanic islands (an island arc system) along the subduction zone. Large bodies of magma can also eventually solidify deep within the crust.

A walk along the ridge of the Malvern Hills is a great way to appreciate how geology affects the development of the landscape. To the east is a broad plain underlain by Triassic sandstones, clays and siltstones, deposited by seasonal lakes, rivers and wind in an arid desert. Just on the horizon one can make out the escarpment that marks the edge of the harder Jurassic limestones of the Cotswolds. To the west is a landscape of hills and vales of mainly Silurian rocks; the harder limestones form the higher ground surrounded by softer sands and clays.

Opposite page: Looking east from the Malvern Hills. The hard igneous rock of Pre-Cambrian age fills the foreground with the broad plain of Triassic sands and clays beyond. Just visible on the horizon is the escarpment of Jurassic limestones, the edge of the Cotswolds.

Left: Looking south towards the southern end of the Malvern Hills. The end part which includes the Herefordshire Beacon has been shifted to the west (right) by a fault caused by tectonic activity during the Variscan Orogeny (see page 74). In fact these Pre-Cambrian rocks were thrust over younger Silurian rocks to the west during the orogeny.

The Early Palaeozoic Era 542-416 million years BP - *the Cambrian, Ordovician and Silurian periods*

The Early Palaeozoic lasted from around 542 to 416 million years ago and is divided into three periods, the Cambrian, Ordovician and Silurian. These names are all associated with Wales, Cambria being the Latin form of the name for Wales while the Ordovices and Silures were both tribes of ancient Britain whose territories were in Wales. It may come as no surprise that rocks from these periods are found in Wales but also extensively in Scotland, the Lake District and the Welsh Borders. There are some outcrops a little further east of these main areas.

Igneous, sedimentary and metamorphic rocks from the Early Palaeozoic tell the story of the creation of a great mountain range, the roots of which can still be seen today. The story begins far south of the equator where the area of crust that was to become much of Scotland was on the southern margin of a continent known as Laurentia. The rest of Britain was further south on the edge of a continent called Avalonia; they were separated by the Iapetus Ocean which was growing ever wider. A variety of sediments were deposited in different environments, some like sandstones and limestones in shallow water on a continental shelf, others like clay in deeper ocean basins. Some sandstone layers were deposited in the deep basins, products of 'turbidity currents', fast moving flows of sediment that tumbled down the slopes from the continental shelf, sometimes triggered by earth movements. Thousands of feet of such sediments accumulated in this manner. By the Late Ordovician and Early Silurian the movement of tectonic plates had changed and the Iapetus Ocean was beginning to close – this would result in the British Isles 'coming together'. The line that marks this coming together is known as the Iapetus Suture and roughly follows the boundary between England and Scotland.

Opposite page: Castlerigg stone circle near Keswick in Cumbria is around 5000 years old. It sits on Ordovician mudstone deposited in a marine environment about 485-465 million years ago. The stones and surrounding hills are volcanic, largely tuff, an explosive product from a volcano. They are also of Ordovician age, about 455-450 million years old, and indicate an active, destructive plate boundary where two oceanic plates were colliding, resulting in an 'island arc' where volcanic islands arose above the descending plate.

Wales

In Cambrian times North Wales was a deep ocean basin where vast amounts of sand and mud were deposited. These sediments were later intensely folded during the Caledonian Orogeny and the closure of the Iapetus Ocean. In the Ordovician there was massive volcanic activity associated with the subduction of the Iapetus. By this time much of South Wales was a shallow shelf sea, bordered by land to the south, but with deeper basins further north.

In the ocean basin of what is now North Wales the sandstones deposited were mainly what are known as turbidites. Earth movements often triggered underwater 'landslides', swirling masses of sediment that had collected on the continental shelf tumbling down the continental slope into the abyss below. These turbidity currents are responsible for depositing much of the sediment that collects in deep ocean basins. As oceans are squeezed out of existence this sediment is crushed and folded into mountain ranges.

Opposite page: The mountain of Cader Idris with its glacier formed cwm below. The mountain is formed of mixed layers of Ordovician volcanic and sedimentary rocks, folded during the Caledonian Orogeny. Notice the strata are tilted (sloping right to left) while the fine grained sedimentary layers show a cleavage dipping steeply left to right; the result of compression in the mountain building process.

Left: Pistyll Cain waterfall in North Wales. The river tumbles over hard Cambrian sandstone layers, turbidites that were deposited in a deep ocean basin. This is on the edge of the 'Harlech Dome' a large, complex anticline, again folded during the Caledonian Orogeny.

The Lower and Middle Cambrian sandstones and mudstones at Caerfai on the Pembrokeshire coast of South Wales were deposited in a shallow sea with highland to the north and south. This area which was to become South Wales was around 60 degrees south of the Equator and plate movement was stretching this part of the crust, eventually creating a deeper ocean. There were a number of rises in sea level during the Cambrian, partly because the Earth was recovering from its "Snowball Earth" phase and melting ice and snow was adding to the oceans but also because plate tectonics was creating more mid-ocean ridges and changing the topography of the sea floor.

The Caerfai Group of sandstones and mudstones are over 550 metres thick, principally derived from the highland to the south. Evidence suggests they were deposited on a delta front and in shallow seas on the northern edge of the continent of Avalonia (see page 62). The land was completely devoid of vegetation at this time, terrestrial plants had not yet evolved. It is not difficult to imagine the erosion of the land surface would have been much faster than it is now.

Also on the Pembrokeshire coast is Marloes Sands. Here sandstones and mudstones from the Lower Silurian are exposed in striking vertical strata deformed in the Caledonian Orogeny. Interbedded with these sedimentary strata are layers of volcanic lavas and tuffs (the Skomer Volcanic Group, named after the nearby island). These represent the youngest volcanic episode associated with the Caledonian Orogeny at a time when the Iapetus Ocean was nearing closure.

Left: Caerbwdy Sandstone is part of the Caerfai Group; it is a durable building stone and was used to construct St David's Cathedral.

Marloes Sands with near vertical layers of sandstones and mudstones from the Silurian Period; folded during the Caledonian Orogeny.

The Lake District

The oldest rocks of the Lake District are the Skiddaw Slates, confusingly not totally slates but also grits and sandstones. They are Ordovician in age and may be up to 20 000 feet thick, having been deposited in quiet, shallow waters of the Iapetus Ocean off the northern coast of Avalonia. As we have seen, slate was originally fine grained sediment and subsequently deformed by the collision of tectonic plates. They form the rounded hills and mountains of Cumbria such as Blencathra. The mountains of the central core of the Lake District are not formed from the slate but from volcanic rocks, the Borrowdale Volcanic Series. As the oceanic plate of the southern Iapetus Ocean slid beneath Avalonia partial melting caused huge volcanic outpourings of lava and ash, the episodes gradually becoming more explosive. Some of the ash fell into the waters of the coastal margin and were later compressed to form the slates like those at Honister, quarried for many years. The 'Borrowdale Volcano' was huge and active for a relatively short time, around 5 million years. It was above sea level and probably a caldera, a vast crater left from the collapse of the roof of a magma chamber. There were many volcanic vents.

At the end of the volcanic episode the landscape began to erode. There were volcanic mountains or perhaps just one; eventually the sea returned covering the now flat land and depositing more sediment. The sea was fairly shallow and limestones and sandstones formed. The southern part of the Lake District is covered by these Silurian sediments. All this time the Iapetus Ocean was gradually shrinking and eventually the Cumbrian rocks would be folded into a great mountain range, the Caledonides (see page 62).

Opposite page: The smooth contours of Blencathra, formed from the Skiddaw Slates.
Right: The Langdales, formed from the Borrowdale Volcanics.

The Caledonian Orogeny

An orogeny is basically a mountain building episode; a series of events spread over millions of years that result in the formation of a mountain range. The end stage is often a collision between two continental plates, continental crust is relatively light and will not subduct under another plate as cold, oceanic crust does. An ocean that originally separated the plates will have already subducted and closed and then vast thicknesses of sediment that once covered the ocean floor will be thrust up into a mountain range. Before that there will have been intense volcanic activity in the vicinity of the descending oceanic crust as magma from the melting plate rises through the crust. This is one of the main mechanisms through which continental crust is formed (see page 26).

The Caledonian Orogeny took place over around 100 million years from roughly 490 – 390 million years BP. It is perhaps misleading to think of it as one event, there were many different phases as the tectonic plates involved moved over the Earth. The movement of plates is a complicated affair, they are moving around on a sphere. There can be phases where a region of crust is being compressed, others where it is being stretched or extended or times when two areas are sliding past each other. These movements over that 100 million years have been responsible for the creation of much of Britain.

At the beginning of the Early Palaeozoic a huge supercontinent named Gondwana straddled the South Pole. On its northern edge a smaller continent named Avalonia began to split off from Gondwana. Far to the north the continent of Laurentia lay the other side of the Iapetus Ocean which was being subducted on both its margins. At this stage England and Wales lay on the northern edge of Avalonia, most of Scotland on Laurentia. Gradually the Iapetus Ocean closed with much volcanic activity associated with its subducting southern margin. Around 420 million years ago England and Scotland were finally joined, roughly along the line of the present boundary between the two. Sediment

Opposite page: Glencoe - these mountains are volcanic, part granitic and other lavas and tuffs that formed in the early Devonian as the two continents of Gondwana and Laurentia collided and joined.

that had collected in various environments throughout the ocean was squeezed and uplifted into a huge mountain chain, the Caledonides. Today remnants of those mountains remain, and remnants of the various episodes that led up to the final mountain building. The volcanic tuffs, ash and lavas that form the beautiful hills of the Lake District derive from the volcanic activity as the oceanic crust of the Iapetus Ocean was being subducted. Many of Scotland's mountains, although still impressive, are the eroded remains of once even mightier peaks, some formed of the granite that once cooled slowly deep inside the orogenic belt. Elsewhere in Scotland and Wales folded and contorted strata can be seen in sediments that were crushed as tectonic plates collided.

Opposite page left: Tightly folded Dalradian (Pre-Cambrian) slates on the island of Kerrera off the west coast of Scotland. These were deeply buried sediments deformed during the Caledonian Orogeny. What were fine grained sediments now show a slaty cleavage (see page 42) while the harder sandstone layers show the tight folding.

Opposite page top right: Remnants of the Caledonian mountains near Ballachulish, Scotland. The peaks are mostly igneous plutons that were once deep inside the mountain range and have been exposed by erosion over millions of years.

Opposite page bottom right: On the peninsular of Ardnamurchan on the west coast of Scotland this rock is known as the Moine schist. Originally a fine to medium grained sediment it has been deformed to a greater extent than slate by the mountain building process. It has a very tight, wavy cleavage and the heat and pressure have resulted in new minerals forming. Its shiny texture is the product of the mineral mica.

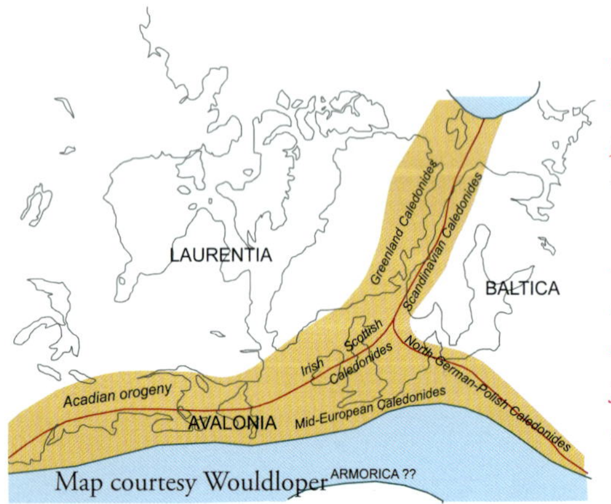

Map courtesy Wouldloper

Left: A diagram showing the rough position of the modern continents as ancient landmasses collided and formed the Caledonides mountains. The red line across Britain shows the rough position of the Iapetus suture where Scotland and England were joined. Note another ancient continent named Baltica was also involved in the collision.

The Late Palaeozoic Era 416 - 251 million years BP - *the Devonian, Carboniferous and Permian periods*

The first period of the Late Palaeozoic is the Devonian. It takes its name from the county where lots of rocks from this age are exposed. Cornwall too has Devonian sedimentary rocks and in both counties they are rocks that were formed in a variety of marine environments. This is very different from the rest of Britain where Devonian sediments were formed in continental environments, typically arid deserts or freshwater lakes and rivers. This gives us a picture of Britain after the Caledonian Orogeny, mostly land with the mighty mountains of the Caledonides to the north but sea to the south in what is now Devon and Cornwall. The mountains of the north were similar to the present day Himalayas, the collision of the three continental plates had led to thickening of the crust. The climate was arid and warm, Britain was about 15 to 20 degrees south of the Equator and plants were rare on land – fungi and lichens had begun to colonise the Earth. Erosion of the mountains was rapid and many of the continental Devonian sediments represent deposits from alluvial fans at the foot of highland and the result of flash floods. These deposits became known as the Old Red Sandstone from the colour derived from the oxidation of iron.

To the south Devon and Cornwall were covered by the northern edge of the Rheic Ocean, and throughout the period various marine sediments were deposited. North Devon was on the very edge and here the sediments were occasionally continental as the sea level fluctuated. Cornwall, on the other hand, saw more deeper water sediments such as shales and mudstones, later metamorphosed to the ubiquitous slates by the Variscan Orogeny.

Opposite page: The isle of Kerrera in North West Scotland; this is Devonian conglomerate sitting unconformably on Pre-Cambrian slate. There are pebbles and much larger rocks of all sizes; this layer has been eroded from the Caledonides mountains and must have been carried by fast flowing water, probably a flash flood.

Inset: Old Red Sandstone layers at Manorbier, South Wales; not just sandstones but also mudstones and siltstones. These beds are upright, indicating they have been folded - this time by a later orogeny, the Variscan (see page 74).

The Lizard Ophiolite Complex

A rather technical sounding name for these rocks of the Lizard peninsular of Cornwall, they represent an uncommon occurrence on the continental crust. They are Devonian in age, dating roughly from about 400 million years ago and were originally part of the oceanic crust and the top of the mantle beneath. Oceanic crust is denser than continental crust and when tectonic plates collide is usually subducted under continental crust. However, young, relatively hot oceanic crust is less dense than older, cooler oceanic crust and when such plates collide has a greater chance of being obducted, or thrust onto adjoining continental crust. This is known as an ophiolite.

Shortly after its formation (geologically speaking) at an oceanic ridge, this bit of oceanic crust found itself caught between converging plates as the ocean was squeezed out of existence. A slice of the ocean crust and part of the mantle beneath was pushed upwards instead of being subducted. It is this oceanic crust and mantle we see today on the Lizard.

The Lizard is known for its spectacular cliffs of serpentinite, a rock formed almost wholly of the mineral serpentine. With its deep reds and greens it polishes well and has spawned a local industry making souvenirs. The serpentinite is, in fact, part of the mantle. As one oceanic plate was subducting fissures allowed water to seep into the mantle above. This caused the minerals olivine and pyroxene, the main constituents of the mantle rock peridotite, to hydrate and become serpentine, also causing it to be less dense and therefore more buoyant. This piece of oceanic lithosphere was then scraped onto the continental crust.

Right: Serpentinite pebbles at Kynance Cove.

Kynance Cove on the Lizard, Cornwall. These cliffs were once part of the Earth's mantle.

The Carboniferous Period

As the Devonian gave way to the Carboniferous Period there was a global rise in sea level, partly associated with the retreat of the massive ice sheet over the continent of Gondwana. Consequently Carboniferous sediments are found extensively over northern England, the South West and South Wales. Those dating from the early part of the period are mainly marine sediments. By the mid-Carboniferous most of England and Wales was ocean apart from land across part of the Midlands and central Wales. There were still mountains in northern Scotland. After the mountain building the thickened crust was relatively buoyant (the Himalayas continue to rise) and the highland areas were eroded quickly with sediment collecting in neighbouring basins. This led to a situation across northern England of buoyant blocks with adjacent subsiding basins where great thicknesses of sediment accumulated. The sea level eventually rose to cover the blocks where thinner sequences of sediment were deposited.

The Early Carboniferous saw the deposition across many parts of England and Wales of the Carboniferous Limestone. Formed in a shallow, tropical sea teeming with life this limestone is associated with much iconic scenery in places such as Yorkshire, South Wales and the Mendip Hills. It contains many fossils, especially corals, brachiopods and molluscs.

All three pictures on these pages show Carboniferous limestone. Left is 'limestone pavement' above Malham Cove, Yorkshire. Rainwater has selectively eroded joints in the rock after ice sheets had eroded any surface soils. Joints are the result of tension caused by uplift, perhaps due to the rebound of the crust after ice sheets melted. Opposite page far right shows the limestone layers at Malham Cove.

Opposite page left shows the Green Bridge of Wales in Pembrokeshire, a natural arch carved out of the gently dipping layers of Carboniferous limestone.

By the mid-Carboniferous the situation was somewhat different. Sea level fluctuated and much of northern England saw large river deltas and low-lying alluvial flats. In Yorkshire sediments from these environments are known as the Yoredale Series, typically bands of hard sandstones, softer shales and limestones when the sea encroached. They form the characteristic stepped sides of the moors and can be seen in the many waterfalls. Great forested swamps began to form, populated by tree ferns, packed closely together and around ten metres tall. Vegetation was dense over much of the world and the oxygen level had risen to around fifty per cent higher than it is today. The trees were unlike those in today's forests; they had hard bark but inside were soft and spongy. The bark looked like scales and the whole trunk could photosynthesise. Their dense, shallow roots intertwined trapping water and decaying organic matter. It is these swamps that are the source of much of the world's coal and they have given the name to this geological period – the Carboniferous.

Opposite page: Mill Gill Force waterfall in the Yorkshire Dales. The layers are rocks from the Yoredale Series, sandstones and shales laid down in river deltas and limestones when the sea transgressed periodically. Notice here and Amroth cliff to the left, the strata are lying horizontally. These strata have not been subject to deformation.

Left: Amroth cliff in Pembrokeshire, South Wales. In the middle of the cliff is a coal seam; the great coal swamps existed here too, south of the landmass that stretched across middle England and middle Wales. Inset is a close up of the coal seam.

The Variscan Orogeny

The second great mountain building episode that has helped 'build' Britain is known as the Variscan Orogeny. An orogeny is not a simple, single event but rather a series of connected episodes over many millions of years, all associated with the movement of tectonic plates and culminating in two coming together. There may be periods where the crust is extending, forming ocean basins where vast quantities of sediment are deposited. There will be periods of volcanic activity and periods where the crust is slowly deformed as slabs of continental crust inch slowly together.

The Variscan Orogeny took place over about 100 million years from roughly 400-300 my BP. The continent of Gondwana moved northwards eventually colliding with a continent named Laurussia, crushing ocean basins and piling up a huge mountain range. Remnants of these mighty mountains include the Black Forest, the Iberian mountains and the

Below left: Highly contorted minor folds at Trevellas Porth, Cornwall. The bands are just a few centimetres thick, some less than a centimetre.
Below right: Tight, chevron folds in the cliff at Crackington Haven, Cornwall.
Opposite page: Folded sedimentary layers on Northcott Beach, Devon. All these folds are associated with the Variscan Orogeny.

Massif Central. In Britain the main remnants are the crushed and deformed sediments in South West England and the great mass of granite that underlies this peninsular, peaks of which have been revealed by erosion. The Devonian slates that form many of Cornwall's impressive cliffs represent the eroded core of part of the great mountain chain – remember, slates are fine grained sediments crushed and baked at great depths.

The rock strata of Devon and Cornwall display some spectacular folds, including large scale structures known as 'nappes' where whole sections of crust have been thrust many kilometres northwards. Such structures cannot be readily identified by casual examination except in young mountain ranges like the Alps where they are sometimes visible on the side of peaks. The Carboniferous and Devonian sediments of South Wales also display folds formed during the Variscan Orogeny but these are less intense than those seen in Devon and Cornwall, indicating that this region was further from the centre of deformation (see page 36).

When the plates had collided and the mountains formed the Earth's continental crust had all coalesced into a supercontinent known as Pangaea which stretched from the South Pole well into the northern hemisphere. By now Britain was land and in tropical latitudes just north of the Equator. The climate was hot and dry and the land mostly desert. Later on shallow, inland seas developed to the north east and north west where deposits of salt formed which are still mined today.

Perhaps obviously, the collision had resulted in a considerable thickening of the crust which then began to collapse under its own weight, sinking deeper into the mantle and causing tension in the upper crust. Huge amounts of sediment that had collected in ocean basins and been crushed and deformed were melted by mantle heat and subsequently cooled over thousands of years into granite, forming what's known as the Cornubian batholith, a huge chamber of granite that sits under South West England. The cover of around two to three kilometres of sedimentary rock has been eroded exposing the tops of the batholith. This erosion allowed the granite to expand causing horizontal joints. This, together with the vertical joints that formed as the granite cooled resulted in the characteristic shapes of the granite tors we see today.

Above left: The Cheesewring, a granite tor near Minions, Cornwall.

Above right: Porthleven, Cornwall: the brownish rock on the right is 'greenstone', originally a 'basic' igneous rock (see page 78) intruded into the Devonian slate on the left. This happened as the continental plates were closing and the oceanic plate was subducting. The junction between the two is vague, indicating the slate may have been an unconsolidated sediment when the intrusion happened. These greenstones were later metamorphosed by heat from the granite intrusions.

Right: A close up of the texture of granite on Bodmin Moor. The oblong crystals are feldspar. These crystallized first from the magma and so were able to grow 'unhindered'. Other minerals like quartz crystallized at a lower temperature and filled in the space between the feldspar. The feldspar crystals seem to be roughly aligned, indicating the magma may have been flowing as it crystallized.

The Whin Sill

One of Britain's most spectacular geological features is the Whin Sill or more accurately the Great Whin Sill Complex which stretches across northern England. Geologically speaking a sill is a usually flat lying sheet of igneous rock that has been intruded into surrounding rocks, often between sedimentary strata. They differ from dykes which are typically intruded across surrounding layers, perhaps following faults or other lines of weakness.

The Whin Sill is a series of sills and some associated dykes all intruded around the same time. It is estimated that the total volume exceeds 200 cubic kilometres on land with much more probably under the sea. The rock of the sill is known as dolerite, a 'basic' igneous rock, which just means it has a relatively low silica content; conversely an 'acid' igneous rock (like granite) has a relatively high silica content; neither have anything to do with pH. Dolerite is a medium grained rock meaning its crystals can be seen with the aid of a hand lens, indicating it cooled relatively quickly but not as quickly as lava extruded on the surface.

It was formed just under 300 million years ago at the end or closing stages of the Variscan Orogeny and at the end of the Carboniferous and beginning of the Permian Period. We have seen that the Carboniferous rocks of South West England were highly deformed by the orogeny but the effects of the mountain building were felt further north too. When a range of fold mountains has been formed the thickened crust can be gravitationally unstable and can lead to the newly formed mountains sinking and spreading. This is known as orogenic collapse and leads to parts of the crust extending, leading to hot magma rising from the mantle below. This is how the magma of the Whin Sill formed.

Opposite page: The Great Whin Sill in Northumberland.
Right: Hadrian's Wall sits on top of the Whin Sill for many miles.

Pangaea and the Permo-Trias

Following the Variscan Orogeny which had finished around 300 million years ago in the Late Carboniferous, the continental landmasses riding on the backs of the tectonic plates had happened to coalesce into one supercontinent geologists have called Pangaea (from Ancient Greek, meaning 'whole earth'). This stretched from the northern polar region to the southern. The climate warmed and most of the supercontinent was a hot, arid desert. This included most of Britain and these conditions continued throughout the Permian and Triassic periods and into the Jurassic when eventually rifting from plate tectonics caused Pangaea to break up.

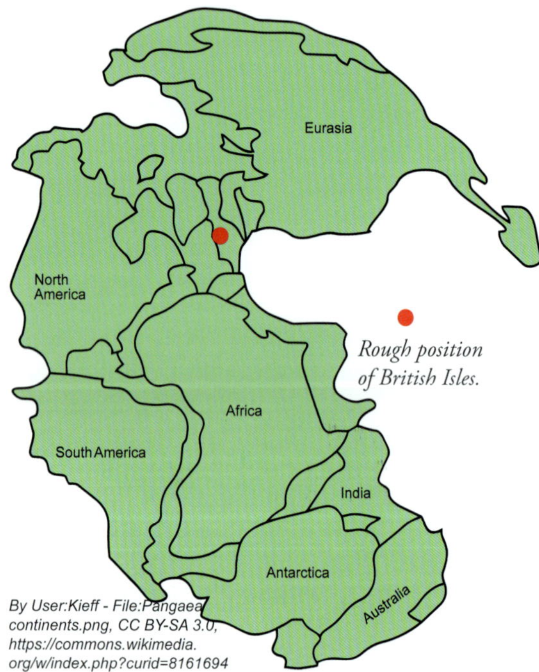

Pangaea was fully formed in the Early Permian and began to break up in the Early Jurassic. Evidence for the existence of this supercontinent is now overwhelming, but at first it was the notion that the shape of the various continents seemed to fit together. Geological evidence now includes the existence of identical fossil species in areas that are now thousands of miles apart and that the Appalachian mountain chain of North America is part of the Caledonian mountain chain whose eroded remnants are seen in Scandinavia, Northern England and Scotland.

Opposite page: Ladram Bay in Devon. The sea stacks and lower cliffs are of the Triassic Otter Sandstone while the change in gradient of the tall cliff reflects the change to the Mercia Mudstone. These are continental deposits laid down in seasonal lakes or by rivers in the arid landscape of the supercontinent of Pangaea.

Left: The map left shows the approximate configuration of the supercontinent of Pangaea.

Permian and Triassic rocks across Britain are what are known as continental deposits, sediments that formed on land by rivers, lakes and the wind. The rivers and lakes were seasonal, much like those that form in arid climates today. Typical sediments are red sandstones, sometimes eroded material from the Variscan mountains, coloured by the oxidation of iron. Deposits of salt formed on the beds of slowly drying lakes and lagoons, the source of the salt that has been mined in Cheshire for centuries. These conditions continued across the Permian and Triassic periods and has led to these two periods being together known as the Permo-Trias.

In the West Midlands the village of Kinver is well-known as the last place in England to see inhabited troglodyte dwellings. These cave houses are cut into the Bridgnorth Sandstone from the Early Permian. They are aeolian or wind blown sandstones with large scale cross (or current) bedding (see picture and caption below) formed as desert dunes migrated across the landscape.

Opposite page: The Holy Austin cave houses at Kinver were cut into the Bridgnorth Sandstone. This is an aeolian or wind blown sandstone; the angled layering represents desert dunes. The current bedding of wind blown dunes is typically on a larger scale than the fluvial current bedding (left).

Left: A close up of the cliff face in Ladram Bay, Devon. At the bottom the sandstone shows 'current bedding', inclined layers laid down by seasonal rivers and reflecting how the river changed course over time. Inclined layers would be laid down on the outside of a meander and as the river changed course layers at a different angle would be laid down above.

HARRY POTTER

The Mesozoic Era 251 - 65 million years BP - *the Triassic, Jurassic and Cretaceous periods*

As we have just seen, the last period of the Late Palaeozoic and the first period of the Mesozoic have been described together as the Permo-Trias since similar conditions existed over both periods and similar sedimentary deposits were formed. Why then are they put in different eras? The reason is that periods and eras are defined by their fossil assemblages; at the end of the Permian was the greatest mass extinction event the Earth has ever known. Popularly called "The Great Dying" with around 80% of marine species perishing and about 70% of land vertebrates. The cause of this is still debated although vast outpourings of lava in certain parts of the world may have been a factor, resulting in raised global temperatures and oxygen starved oceans.

In the Jurassic Period rifting began to break up the supercontinent leading to new seas and oceans forming. This created new environments and life soon began to flourish again. A wonderful variety of species evolved, including, famously, the dinosaurs, giant marine reptiles and many varied species of ammonites. A record of this fauna is preserved in the rocks of the Jurassic Coast of East Devon and Dorset and is the main reason for it being awarded World Heritage status. The Jurassic seas spread over much of Britain, at times depositing clays and sandstones which underlie fertile vales and at others thick layers of wonderful, golden coloured limestone, typically forming higher ground and escarpments. The climate was warm and sub-tropical and near the coast vast, swampy forests were home to giant, plant eating dinosaurs.

By the Mid Jurassic the sea level was high and much of southern and middle England was under the sea. The South West and Wales was still high ground as was much of northern England and Scotland. Towards the end of the Jurassic the sea level fell and in Dorset the famous Portland Stone was deposited in very shallow shelf seas while the Purbeck Stone that followed was formed in coastal lagoons, often covered in swampy forests where dinosaurs roamed.

Picture opposite: Lower Jurassic strata at Lyme Regis, Dorset. These alternating bands of limestones and clays were formed in a shallow sea as Pangaea began to break up. The sea teemed with life, note the large ammonite in the foreground.

The Jurassic Period

The Jurassic Period is named after the Jura mountains in France where these rocks were first studied. At this time the Earth was experiencing a warm, humid climate - there were no ice caps. As Pangaea began to break up and new oceans were created, the changing undersea topography caused by sea-floor spreading led to a global rise in sea level. In the Early Jurassic most of England and Wales was under the Tethys Ocean and marine sedimentation continued across much of England for the rest of the period but with landmasses across much of Wales, Scotland and the South West.

There had been another mass extinction event at the end of the Triassic Period but life recovered and flourished once more in the warm, shallow seas of the Jurassic. This is the age of giant marine reptiles, numerous ammonite species and, of course, dinosaurs on land. A wonderful fossil record of this fauna is found in the rocks of the Jurassic Coast of East Devon and Dorset, as well as other locations such as the Somerset and Yorkshire coasts, the road cuttings and quarries of the Cotswolds.

The explosion of life led to the formation of the wonderful golden limestones of Dorset, Somerset and the Cotswolds. At other times different conditions resulted in the deposition of clays and sands. Today the limestones form relatively high ground with escarpments that overlook fertile clay vales. The marine Oxford Clay deposited around 160 million years ago covers much of South East England, parts of which are also rich in fossils.

Pictures opposite clockwise from top left: 1. The magnificent fan vaulting in the ceiling of Sherborne Abbey, Dorset. Carved in beautiful, golden Ham Hill Stone, a Jurassic limestone deposited in a warm, shallow sea. Shell and coral fragments collected in the sand and clay on the sea bed.
2. The Blackmore Vale, Dorset, a rich agricultural land of Jurassic clays and sands.
3. Broadway, Worcestershire. This famously picturesque village is built almost entirely of golden, Jurassic limestone.
4. Portland Bill, Dorset. This limestone from the Upper Jurassic was again formed in a warm, shallow sea. Often called the best building stone in the world.

The Cretaceous Period

The Cretaceous Period is named after the Latin 'creta' meaning 'chalk'. At around 79 million years it is a relatively long period. At the beginning most of northern Britain, Wales and South West England was land. Sediments known as the Wealden Group, mainly sands and clays, were deposited on river floodplains, in coastal lagoons and river deltas. These are found in the South East and Dorset.

Sea levels gradually rose and eventually most of Britain was covered by sea and the deposition began of a unique rock - the Chalk. The purest Chalk is almost entirely organic, composed of the microscopic calcite armour of coccolithophores, tiny single celled organisms that produce calcium carbonate shields to protect themselves. Each one is beautifully intricate and the organism sheds them periodically which then slowly sink to the bottom of the ocean, eventually building up a thick calcareous ooze on the sea floor. It is this which eventually turned to chalk.

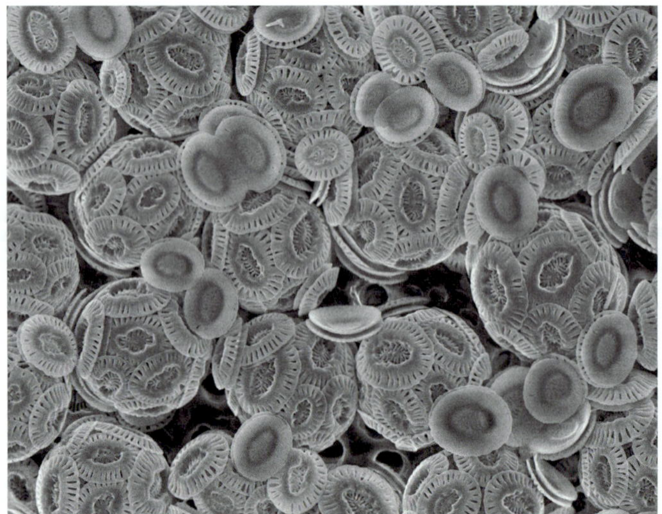

The conditions in which the Chalk was deposited remain something of a mystery. As noted, the purest Chalk is almost entirely organic, there is little or no terrigenous material, sand or clay. This is very unusual for a marine environment and it has been speculated that perhaps the Chalk was deposited in a shallow sea surrounded by waterless desert to which no rivers brought sediment.

Left: Single celled coccolithophores surrounded by their intricate calcite shields. Each shield is roughly between 2 and 25 micrometres across (a micrometre is one millionth of a metre). Courtesy Dr Alison Taylor, Wikimedia Commons.
Right: The magnificent Chalk coastline looking east from White Nothe on the Jurassic Coast of Dorset.

The Cenozoic Era 66 million years BP to today - *the Palaeogene, Neogene and Quaternary periods*

Towards the end of the Cretaceous the Chalk sea covered practically all of Britain but by the early Palaeogene, the next geological period after the major extinction, most of the country had been uplifted, the only areas remaining under the sea were the Hampshire Basin and the London Basin. It has been suggested that this uplift may have been due to the collision of the African and European plates and the onset of the Alpine mountain building episode but some research has suggested it was related to an uprising mantle plume to the north. Whatever the reason, sediments from the Cenozoic Era are largely confined to central southern England and eastern southern England.

However, this does not mean that Cenozoic rocks are not found elsewhere in Britain. There were vast outpourings of lava and other volcanic activity in the Palaeocene and early Eocene epochs (the first two subdivisions of the Palaeogene Period) particularly in North West Scotland. This was associated with the early stages of the North Atlantic opening as two plates diverged. This 'extensional tectonics' caused a release of pressure in the upper mantle resulting in the formation of huge amounts of magma. Vast amounts of basaltic lava erupted and a chain of volcanoes formed. The remains of this activity can be seen over much of North West Scotland, forming some magnificent scenery.

Left: A basalt or dolerite dyke near Port Appin in Argyll and Bute, North West Scotland. This was part of the volcanic activity associated with the opening of the Atlantic Ocean. The dyke is intruded into ancient metamorphosed sediments of Pre-Cambrian age.

Opposite page: The promontory of Stallachan Dubha with the peak of Ben Hiant behind. This is on the peninsular of Ardnamurchan, again in Argyll and Bute. Ben Hiant has a complex geology, being formed of a number of volcanic rocks, emanating from a huge volcano, that flowed over the ancient rocks known as the Moine Schist, again of Pre-Cambrian age. The so-called 'Mull Volcano' is thought to have been situated about 25 km to the south.

Above left and right: The island of Staffa and Fingal's Cave (right), off North West Scotland, showing the columnar basalt with 'slaggy' basalt on top erupted from fissures in the newly developing Atlantic Ocean as tectonic plates slowly drifted apart. The columns form as the lava rapidly cools and contracts. Similar structures are seen as desiccation cracks when lake beds dry out.

Bottom left: The mountain of Ben More on the island of Mull, North West Scotland. The island is mostly basalt lava, again erupted from fissures. It has been shaped by glaciation.

Opposite page: Looking from the Ardnamurchan peninsular, North West Scotland to the island of Eigg. The dramatic ridge (middle picture) is formed of thick, 'acidic' lava which has withstood weathering better than the basalt lavas. Beyond are the mighty Cuillin mountains of Skye, hard, intrusive igneous rocks.

The Hampshire and London Basins

The Palaeogene sediments of East Dorset, Hampshire and the Isle of Wight form part of the Hampshire Basin, surrounded to the north by the Cretaceous Chalk. They are mainly unconsolidated clastic (broken pieces of older rocks) sediments, clays, silts and sandstones with a few limestones. They were deposited in shallow seas and in deltas, estuaries, rivers and lakes on the margins of land that bordered the North Sea. This variety of sedimentary environments indicates a sea level that was changing fairly rapidly in geological terms. Climate changes played a part in this; during the Paleocene and Early Eocene epochs Britain experienced a severe 'hothouse' climate, ten to twenty degrees Celsius hotter than today, even though Britain was around 40^0 north of the Equator. Sea levels were around ninety metres higher than today's levels.

Tectonics also played a part in sea level change, the African plate was steadily moving towards Europe, culminating in the formation of the Alpine mountain range. The rocks of the Hampshire Basin, the London Basin and the Weald were gently folded by these tectonics during the Miocene Epoch, the first of the Neogene Period.

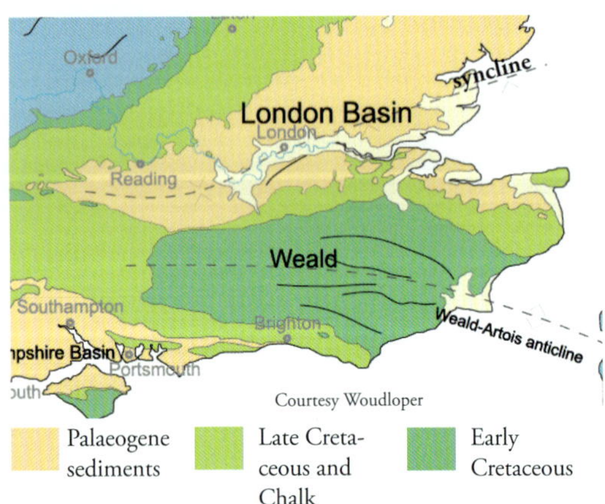

Much of the London Basin is underlain by the London Clay. This blue/grey, often calcareous clay yields many fossils indicating an environment rich in life. It was deposited in a shallow to fairly deep sea and is up to 150 metres thick in places. It has long been used for making bricks and, thankfully for London, is particularly good for tunneling.

Left: A map showing the distribution of Cenozoic sediments. Notice how the Late Cretaceous and Chalk strata occur north and south of the London Basin; all are folded into a gentle syncline. To the south the older Cretaceous sediments are between younger Cretaceous strata; they form part of a gentle anticline (see page 34).

Thaxted, Essex is underlain by the London Clay which dates back to the Eocene Epoch around 50 million years ago. It rests on the Chalk and Late Cretaceous sediments which have been folded into a gentle bowl or syncline. However, it's not the London Clay you find on the surface here but glacial till, sometimes called "boulder clay" dropped by ice sheets during the Anglian Glaciation around half a million years ago. This fertile soil contains many chalk and flint fragments ripped by the ice sheets from the Chalk to the north (see page 96). The largely flat, open landscape makes it ideal for windmills.

The Ice Ages

Britain has been subject to several 'ice ages' over the last half million years. At times much of Britain was covered in ice, probably reaching as far south as London. In highland areas glaciers flowed down pre-existing river valleys while lowland areas were covered by ice sheets, at times hundreds of metres thick. The last ice age or glacial period only ended about 10 000 years ago and it is this episode that has left the most conspicuous features on our landscape.

Basically, the effects of ice sheets and glaciers can be described as either erosional or depositional. Ice will flow down a gradient, albeit quite slowly, and it will gouge the floors and sides of valleys and smooth exposed rock surfaces. It will carry along the eroded material until it stops or slows down when it will deposit this material, usually known as moraine. The effects of glacial erosion and deposition can be seen in all Britain's highland areas, while other areas may predominantly show the depositional effects. Even parts which were not covered by ice have features associated with glacial periods. Southern Britain would have been an area of permafrost where fast flowing meltwater rivers would have flowed during warmer interglacial periods.

Opposite page: This is Ribblesdale in the Yorkshire Dales. A glacier passed through this Dale; material it dropped was smoothed by the glacier into these elongated egg-shaped mounds known as 'drumlins'. The axes of the mounds indicate the orientation of the glacier with the more pointed ends thought to show the direction of flow.

Left: Again in the Yorkshire Dales, this time Swaledale, a wide U-shaped valley cut by a glacier with a 'misfit' river that seems too small.

Top left: The Welsh mountain Cader Idris with 'corrie' or 'cwm' below (see page 57) and with rock smoothed by the flow of a glacier.
Top right: Restronguet Creek, Cornwall: After the Ice Ages sea levels rose from the melting ice and drowned river valleys to form 'rias'. There are many of these in Cornwall.
Bottom left: A block of Jurassic limestone lying by a farm in Thaxted, Essex. No Jurassic sediments outcrop anywhere near here, this is a glacial 'erratic' a block eroded then carried away by a glacier or ice sheet and subsequently dropped as the ice retreated. The front cover shows such erratics in the Yorkshire Dales.
Opposite page: Borrowdale in the Lake District, a classic U-shaped valley cut by a glacier. The sides have been weathered resulting in the formation of scree slopes, now mostly covered by vegetation. The foreground rock shows layers of volcanic ash and tuff and has been scoured and smoothed by the glacier that passed down this valley.

Fossils

Fossils are the remains left in the rocks by living organisms. They may be actually the original material, shell, bone, plant tissue or perhaps original hard parts that have been replaced by another mineral. They may simply be moulds or casts of animals or plants or even traces like worm burrows or footprints. Very rarely, soft tissue may have left an impression in rocks. All these, what we might call 'obvious fossils', date from the Cambrian Period onwards, although micro-fossils and other tell-tale signs of life are now recognised in rocks over three and a half billion years old.

Fossils from the Cambrian to the Cenozoic Era can be found in Britain's sedimentary rocks and guides about where to look and collect can be found online, in books and in local heritage centres and museums. Not all strata are fossiliferous, you would be very lucky to find anything in a desert sandstone. It is layers that have been deposited in shallow shelf seas that are more likely to contain fossil remains.

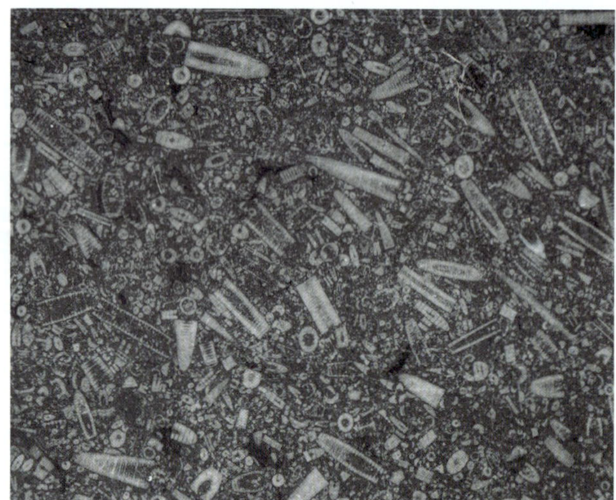

Opposite page: The ammonite pavement on Monmouth Beach, Lyme Regis, Dorset. This limestone layer has dozens of ammonite fossils; sometimes referred to as an 'ammonite graveyard', it may be that a sudden event caused the death of many of these creatures.

Left: This is a tile on the floor of Dent church in the Yorkshire Dales; it contains hundreds of fossils of crinoids, also known as sea lilies, not plants but animals (echinoderms). They were attached by stalks to the sea bed; it is these we see in the tile. Another example of how you can 'collect' fossils with your camera. There are also many museums where you can appreciate the range of fossils found in Britain. The small picture above is an Ichthyosaur skull, courtesy Lyme Regis Museum.

Diversity

Overwhelmingly, diversity is a good thing. Nature shows us that time and time again; it's almost a defining feature of our wonderful planet. All the natural elements formed in the fiery furnaces of long dead stars are present in our rocky globe, although only eight of them make up about 99.5% of the Earth's crust – oxygen, silicon, aluminium, iron, calcium, sodium, potassium and magnesium. These elements combine to form a diverse array of minerals that form our rocks, together with the less abundant carbon which helps form limestones and is, of course, essential for life. The much rarer, heavier, radioactive minerals like uranium play a vital part too; their decay deep within the Earth keeps the interior hot and powers the movement of tectonic plates.

Diversity is essential for a healthy ecosystem and the variety of environments and landscapes created by geological and atmospheric processes has enabled the profusion of life that has existed for many millions of years. In Britain we are perhaps uniquely lucky that our relatively small land area showcases a tremendous geological diversity. We have areas that were once in the core of great mountain ranges, rocks formed by volcanic island arcs, sediments formed in tropical seas and others in arid deserts, lavas and other deposits spewed from giant volcanoes and granites that cooled deep down underneath mountain chains. We even have remnants of the Earth's mantle, the layer beneath the crust.

Many of our sediments yield numerous fossils, in particular the Jurassic Coast was granted World Heritage status partly because its fossil record documents the incredible diversity of life that flourished after the great Permian extinction and before the extinction at the end of the Cretaceous Period.

Finally, our landscape has been shaped by glaciers and ice sheets, relatively recently geologically speaking. The end result is a tremendously varied landscape that draws thousands of tourists to its diverse regions. A little understanding of the many processes and contingencies that have shaped it can only, surely, add to our enjoyment of it; but perhaps more importantly it should foster care and respect for it.

Sandymouth, Cornwall - Carboniferous sandstone ~315 my.

Malvern Hill, Worcestershire - Pre-Cambrian igneous ~850-540 my.

Waterfall at Keld, Yorkshire - Carboniferous Yoredale Series ~340 to 320 my.

Rhosilli, Pembrokeshire - Devonian Old Red Sandstone ~410 my.

Kimmeridge Bay, Dorset - Jurassic Kimmeridge Clay ~ 155my.

Appin, Argyll and Bute, Scotland, Appin Quartzite, Pre-Cambrian ~800-1000 my.

Tintagel, Cornwall, Devonian slate with boudinage ~ 360 my.

Watchet, Somerset - Jurassic limestone and mudstone ~ 200 my.

Longsleddale, Cumbria - Ordovician rhyolite lava ~450 my.

Suggestions for further reading:

Timefulness - Marcia Bjornerud, Princeton University Press.
Turning to Stone - Marcia Bjornerud, Wildfire.
The Planet in a Pebble - Jan Zalasiewicz, Oxford University Press.
How to Read a Rock - Jan Zalasiewicz, The History Press.
The Hidden Landscape - Richard Fortey, Bodley Head.

Acknowledgements:
Many thanks to my wife 'Steve' for much help, support and encouragement.

Above: Tilted layers of Silurian limestone at the Wren's Nest, Dudley, West Midlands. The Silurian rocks here are an inlier, surrounded by Carboniferous sediments. This famous site in the Black Country was the UK's first geological National Nature Reserve. Hundreds of different fossil species, including many trilobites, are found here, documenting life in the shallow, tropical Silurian sea.

Opposite page: Around 55-60 million years ago a column of magma (a mantle plume) began to rise underneath what is now western Scotland. Centred around the island of Mull the bulging crust cracked and fissured spewing great volumes of lava onto the surface and injecting lava into lines of weakness in the crust. These dykes can be seen in many places, this one on the island of Kerrera is my favourite. It makes a perfect spot for sheep to shelter from sun, wind and rain - the latter two possibly being more common!